# BOOTS, SUITS, AND STARTUPS

## Building a Business While in Uniform

# THOMAS KELSEY

Thomas Kelsey -- 1st ed.
Chief Editor, Shannon Buritz
ISBN: 978-1-954757-52-3

The Publisher has strived to be as accurate and complete as possible in the creation of this book.

This book is not intended for use as a legal, business, accounting, or financial advice source. All readers are advised to seek the services of competent professionals in legal, business, accounting, and finance fields.

Like anything else in life, there are no guarantees of income or results in practical advice books. Readers are cautioned to rely on their judgment about their individual circumstances to act accordingly.

While all attempts have been made to verify the information provided in this publication, the Publisher assumes no responsibility for errors, omissions, or contrary interpretations of the subject matter herein. Any perceived slights of specific persons, peoples, or organizations are unintentional.

*I want to thank my wife, Ashley, and our children for their continued support, motivation, and love. Ashley, our successes would not have been possible without your strength and willingness to walk side by side into the unknown.*

*Also, I would like to thank those who helped me put this book together, specifically Chef Robert Irvine, Scott Stalker, Ernie Hansen, and Ferguson Dale. Additionally, Tim Krause and Michael D. Mitchell, thank you for helping keep our military entrepreneurs out of trouble. Matthew and Kailie Bloomfield, thank you for the many hours spent helping review chapter after chapter.*

*Lastly, I would like to give special acknowledgment to Shannon Buritz, who had the daunting task of making my words legible, and Mark Imperial, who pulled this entire project together.*

*"Turn your military discipline and forward thinking into business success. Stay focused, face challenges head-on, and transform your goals into thriving ventures. Your military skills can lead you to remarkable achievements."*

—Ernie Hansen, Co-Owner, Media Shield

~

*"Military service and entrepreneurship are two of the most demanding pursuits in the world. I would know because I've done them both, and so has Thomas Kelsey. So, if you're going to put the world on your shoulders and take on both of these things at the same time, wouldn't you want to take advice from someone who's walked the walk? I know I would, and I know you're in great hands with Kelsey and this book. It is everything you'd want in such a book—all at once instructive, inspiring, and actionable. Put its advice to use and begin to watch your wildest dreams come true."*

—Robert Irvine, Chef, TV Host,
Author, Philanthropist, and Honorary
Chief Petty Officer (U.S. Navy)

# CONTENTS

# INTRODUCTION

## The Military Entrepreneur: Redefining Success While on Active Duty

The military has trained you to plan, strategize, and execute. You've spent years leading people, tackling challenges head-on, and, most importantly, serving your country. Now, you're exploring a new mission—the world of business.

My name is Thomas Kelsey. I've walked in your boots, and now I'm here to share my experiences creating a successful business while on active duty. Together we will delve into the process, the rules, the strategies, and the resources. We will uncover how this transition can benefit you, your family, and our military community.

When we talk about business success, there is no one-size-fits-all formula. That's especially true for active-duty military personnel who not only grapple with typical entrepreneurial challenges but also have unique circumstances to navigate. I aim to share my understanding of how you can create and sustain a successful business while still serving our nation.

Some view the military as having a ceiling on the money you can make or invest towards retirement based on its defined pay grades, promotion requirements, and rigorous scheduling because of training and deployments. The military does not need to be a barrier, though, and I have built my entrepreneurship foundation specifically because of military service. The goal of this book is to guide those individuals who are looking to combine military duty with business ownership.

The term "military entrepreneurship" might be unfamiliar to some, but it represents a growing segment of individuals leveraging their military training to create successful ventures while still on active duty. This trend builds upon the solid foundation of veteran entrepreneurship, but there's a unique twist: the endeavor starts while serving.

The Franchise Business Review found that veteran franchises are about 65% more successful than their civilian counterparts, thanks to their military training. The military teaches leadership, perseverance, adaptability, planning, and the ability to work with diverse groups – that can be directly applied to business ownership. You have unique skills and career training that, when utilized properly, can become the basis for a successful business venture.

You may not realize how valuable your planning and risk management skills are in a business setting. In military operations, planning involves establishing a clear end goal and formulating a strategy to achieve it. This skill becomes crucial when setting up a business, allowing military entrepreneurs to visualize their objectives and the steps needed to reach them.

The ability to work with diverse groups of people is another valuable skill embraced by the military. As a melting pot, the military has individuals from varied backgrounds and cultures, fostering an environment that encourages collaboration and mutual reliance. This prepares service men and women for the civilian world, where your business's success or failure and professional relationships will reflect the morals you portray.

## Possible Pathways After the Military

Post-military careers are diverse, with many continuing to serve the government in different capacities while others transition their skills into civilian sectors. Some may leverage their non-technical skills into a completely different path or capitalize on the invaluable soft skills they've gained during their service. However, the unique aspect of this book is that it encourages active-duty individuals to consider business ownership as a viable path – not only as a source of additional income but also as a way to create a legacy and secure their future.

## Fears and Misconceptions

Military entrepreneurs do face specific unique challenges. One is the lack of control over their time, given the nature of their service, which is prioritized over all else. Another is the fear of failure, an inherent risk in starting a business. While valid, these fears can be managed with appropriate planning and a supportive network.

Some military personnel may harbor misconceptions about entrepreneurship, like the belief that certain business ventures are off-limits due to potential conflicts of interest. As we will see, these concerns can often be addressed through clear communication and a thorough understanding of military policies and regulations.

## The Value of Success

Your definition of success is unique to you and could be long-term business growth, the need to replace an income, or even helping your spouse take control of a portion of their ordinarily hectic life. The path to success may not be straightforward. Many fear that the military structure may not support business ownership due to potential conflicts of interest and time commitments. This book will dispel your fears and misconceptions about entrepreneurship while on active duty.

If you are in the military and considering entrepreneurship, you probably have many questions, such as:

- "Am I allowed to do this while still on active duty?"
- "Is it possible to start a business without being 'all in' at all times?"
- "How do I manage my time commitment?"
- "What processes and legal documents do I need?"
- "What resources are available?"

The following chapters will answer all these questions and more while exploring various topics:

THOMAS KELSEY

- Why business ownership is a viable option during active service
- How to make your venture legal
- Creating a plan for success
- Finding and evaluating resources specific to your needs
- Benefits of entrepreneurship to the military community

As business brokers, my wife and I have firsthand experience managing businesses while on active duty. We have grown our net worth through multiple ventures while serving our country. Our experiences, challenges, and successes have inspired us to share our knowledge and help others in the military community explore the vast potential of business ownership.

With this book's strategies, insights, and guidance, I hope to help you, the active-duty military entrepreneur, craft your path to success. It's not just possible; it's a journey worth embarking on.

Let's get started.

- *Thomas Kelsey*

# CHAPTER ONE

## Military Commitments Meet Entrepreneurial Dreams

———

This chapter offers a glimpse into the core purpose of this book and touches upon the unique benefits and challenges you might face while balancing military service and business ownership. You might wonder, "What kind of business can I start that won't clash with my hectic military schedule?" From my experience and observations, real estate often stands out. It's more about passive investing initially, allowing you to transform it into a full-fledged business over time gradually. Apart from that, other ventures that have seen success include professional services, content creation, and e-commerce. These sectors allow you to apply your skills and grow but have the flexibility to mold business hours around your service commitments.

## Striking the Balance Between Military Duties and Entrepreneurship

For every military entrepreneur, it's fundamental to acknowledge one undeniable truth: Your military commitment always takes precedence. It's a bond, sealed when you sign on that dotted line. However, while this unwavering dedication is a given, it can pose difficulties, especially if you stretch yourself too thin financially in your business. Overburdening yourself with hefty loans or financial commitments in business can create undue stress and jeopardize your military obligations.

So, how do you juggle the two? The answer lies in meticulous scheduling, leaning on automation tools, and often burning the midnight oil to ensure your business thrives. You must have a supportive team that understands the nuances of your military role and assists you in managing the business during your absence. We'll go deeper into building such teams in a later chapter, but for now, let's look at one example of striking a successful balance.

As I mentioned at the start of the chapter, real estate investing is a popular choice amongst military entrepreneurs. At first glance, real estate may seem far removed from the conventional business arena, especially in the initial stages of investment. However, as the journey progresses, the nuances of business management become more apparent.

Imagine being an active-duty military entrepreneur who has just ventured into real estate, acquiring a single property. The thrill of this new investment is palpable, but so is an unexpected PCS or deployment. This predicament can pose a significant challenge:

How do you manage your property while fulfilling your military duties?

The solution? Outsourcing and delegation.

By hiring a property manager, military real estate investors can ensure their property is well-maintained and tenants are looked after during their absence. This provides the smooth running of their real estate business and introduces them to delegating tasks. It's a critical step in scaling any business – learning to trust professionals, property managers, plumbers, or other specialists to manage various aspects of your venture.

The realm of real estate offers a unique business perspective for military entrepreneurs. It's a world where assets are tangible, and the importance of delegation becomes evident quickly. By building trust with professionals and embracing the power of outsourcing, military entrepreneurs can carve a successful path in real estate, ensuring their property and military commitments are well-balanced.

### Finding the Right Business for You

It's important to align your business choice with your skills and interests. One of the best ways is through professional advice. Most people first venturing into the business world are slammed with countless resources and a plethora of information. The landscape is full of known-knowns, known-unknowns, and unknown-unknowns. This information overload and desire to learn as much as possible often leaves people not sure what they're looking for but with a clearer idea of what they don't want. That's where an expert steps in. A business advisor can guide you towards opportunities that resonate with your abilities by helping you shape a vision for your business that matches your goals. It's reminiscent of impactful chats with a college professor – offering guidance, mentoring, and opening doors you didn't even know existed.

## My Learning Opportunities

You might be curious about my journey – after all, we're embarking on this learning adventure together. So, let me take you back to the beginning.

My initial step into business was sparked by a desire to understand its intricacies, and I decided to obtain a real estate license. I wanted to understand people better, and sales seemed the perfect avenue. In my then somewhat naive view, real estate appeared to be a venture I could manage part-time. Here's a piece of wisdom for you – hardly any business initially thrives on a part-time commitment. It demands a full-time dedication. However, the key lies in optimizing this commitment, ensuring that every moment invested yields the best results. Real estate was not just about property transactions; it was my gateway into business. During my professional growth in this sector, I stumbled upon business brokerage.

My passion for understanding business continued, and I figured I needed a team. I teamed up with a childhood friend, and we were full of ambition! This entrepreneurial spirit led us to kickstart an internet radio business that did some marketing and had a few apps. But our journey panned out differently than planned. While many might label it a failure, I consider it a "budget-friendly learning experience." Real-world experiences teach you things often overlooked or forgotten in a classroom setting. Our biggest pitfall was a lack of clear strategy and commitment to execute that plan. It felt like being in a pinball machine, bouncing from one idea to another, lacking a clear roadmap.

My wife and I wanted to elevate our game and create our legacy. We wanted to establish a genuine business. Franchises caught our attention – their allure lay in the support, training, and tried-and-tested processes, albeit not all franchisors are created equal. After numerous conversations with advisors and franchise companies, we settled on and were approved to become franchisees in the business brokerage industry. It is often overlooked that franchisors also validate your ability to succeed with their brands, and the active-duty aspect can play a role in that process.

Even with a franchise model, training, and support, the initial stages of a business are challenging. There's an inherent learning curve, and as you're pushing to scale and establish, it demands every ounce of your energy. I recommend taking a block of leave when your business launches to ensure you have the time needed to smooth out the initial unexpected hiccups. More dimensions added to our equation were the dynamics of working alongside a spouse and being on active duty. While we were gearing up our business, it was vital to remember that our relationship needed nurturing, too. During the business hustle, we occasionally had to hit the pause button, take a breather, and recenter our focus on what truly mattered.

Building a business is as much about understanding market dynamics as it is about recognizing your personal relationships and needs. It's not about avoiding mistakes but learning from them, evolving, and using those lessons as stepping stones to future success.

## Evaluating Business Ideas

When military personnel approach me, often brimming with potential business ideas, they seek guidance on the best way forward. Here's my advice:

- **Financial Planning:** Gauge what you have available to invest. Understand your financial capacity, whether it's liquid assets, loans, or other financial support.
- **Time Commitment:** Be realistic about the hours you can dedicate. Is it a regular nine-to-five? Weekends? Evenings?
- **Family Support:** Military life already has its set of conflicts. Introducing a business into the mix will affect your family dynamics. Make sure you have their unwavering support.
- **Station Tenure:** In the military, you often hop from one base to another. Consider the longevity of your current location, especially if you're considering a brick-and-mortar establishment.
- **The "Why:"** Your driving force behind starting the business should be crystal clear. Is it philanthropic? Financial independence? Acquiring new skills? This "why" will anchor your journey.

## Existing Business, Startup, or Franchise?

If you find yourself at the crossroads of buying an existing business or building one from the ground up, there are several things to consider. From my perspective and expertise as a business intermediary, the allure of an established track record cannot be underestimated. It's no secret that first-time businesses face steep odds, especially if spearheaded by those without prior startup success.

Opting for a proven or pre-existing business has its benefits. The chance to inherit an established client base, trusted vendors, and a team of skilled employees who know the ropes is invaluable. They bring expertise and ensure a smoother transition as you steer the ship forward.

Every business choice, starting anew, buying an existing venture, or diving into franchising, carries its set of rewards and obstacles. The key lies in aligning your choice with your circumstances, resources, and long-term vision.

### Breathing New Life Into an Established Business

My wife assisted in selling a painting company, a firm that had painted its legacy for over 40 years. The new proprietor walked into a business with an impressive four-decade reputation. While it might sound like stepping into big shoes, this new owner harnessed the company's existing prowess and sprinkled his innovative magic. He collaborated with the seller, amplified the staff by a staggering 300%, and his strategic vision breathed fresh life into the company. They expanded their offerings from painting to now include flooring, carpeting, and more. The narrative here is clear: When you merge the foundations of an established business with fresh perspective and innovation, the sky's the limit!

## Startup

Creating a startup might be a good option if you desire complete control or have a revolutionary idea. It is important to remember that startups have a lower success rate than the other two methods, as the founders are often undercapitalized and forced to handle every aspect of the business. The key is to understand that most decisions in business require a balance of control and resources. Well-known names such as Steve Jobs, Bill Gates, and Mark Cuban are a few of the countless examples of successful companies that were started with minimal capital and an idea. These previous examples of success can be great learning opportunities, but don't be discouraged if your attempts fail and you are forced to pivot. We often focus on the tiny percentage of success and overlook the number of learning opportunities along that journey.

## Franchise

Franchises have caught the eye of many budding entrepreneurs. Let's discuss what this involves, particularly for someone wearing a military uniform. Acquiring an existing business and opting for a franchise usually require a more substantial initial investment than starting from scratch. With franchises, even though you're replicating a proven model, you're essentially introducing it to your local community, which can be overwhelming. Always tread cautiously. Some franchises may have clauses restricting you from other full-time commitments. Given the demanding nature of military service, you must meticulously sift through the details, evaluating each franchise opportunity.

One of the chief attributes of franchising is the structured training and standards set by the franchisor. The initial outlay, coupled with the time commitment for training, is something to be mindful of. Yet, a silver lining exists. These franchise companies have a vested interest in your success. While the initial fee is a part of their revenue, most companies' primary earnings come from the ongoing royalties. Their success is intertwined with yours.

When my wife and I embarked on the franchise journey, we had a blueprint that guided us at every turn. Their promise of a proven track record materialized for us. The franchise model laid out tasks in sequence – Day 1, do this; Day 2, follow that. It's this detailing that ushered us into success. By adhering to the well-charted path, we found ourselves flourishing and thriving.

### Conquering Unique Military Business Challenges

When military personnel take the entrepreneurial plunge, the problems they face are, at times, starkly different from those of civilians. Often, the desire is to transition military experiences into the business realm seamlessly. This is especially common among technical experts and senior members. But it's a whole different ball game for those on active duty.

Potential conflicts of interest exist, especially if one aims to leverage military experience to craft solutions the government might procure. Seeking legal counsel within the military can help mitigate these issues and keep all endeavors above board.

Frequent relocations, deployments, and training events can disrupt business consistency. There's also the need to seek the commander's nod before entering any external ventures. The civilian world offers a bit more autonomy, but military entrepreneurs must always remain within the defined boundaries.

**Tips for Success:**

- **Engage with the Command Team and JAG:** Your command team and Judge Advocate General (JAG) are pivotal. While each commander might have personal perspectives on active-duty personnel running businesses, presenting a well-structured business plan or argument can alleviate concerns.
- **Re-communicate upon Relocation:** If you find yourself relocating, remember you're starting fresh in a new setting. Communicate your business pursuits to your new commander to keep them abreast of your commitments.
- **Plan for Deployments:** Deployments often pull you away for extended periods. The answer to this problem is a stellar team. Hire a reliable squad to manage operations in your absence, mitigate risks, and ensure business continuity.

The journey of a military entrepreneur is filled with distinct challenges. But with proper planning, support structures, and a clear vision, you are well on your way to entrepreneurial success.

# KEY TAKEAWAYS

- For military entrepreneurs, real estate offers a blend of tangible assets and passive investing. With the help of trusted professionals, it's possible to maintain property investments even during deployments, emphasizing the importance of delegation.

- Military obligations always take precedence, but the key to balancing service and business lies in meticulous scheduling, leveraging automation tools, and building a supportive team that understands the dynamics of both worlds.

- Whether considering starting fresh, acquiring an existing venture, or embracing franchising, always align the choice with personal circumstances, resources, and a clear vision for the future.

- Engage with your command team and the JAG to keep your business pursuits within permissible boundaries. Consistent communication, especially during relocations and deployments, is essential to maintain transparency and keep the business operating smoothly.

- The journey of a military entrepreneur is uniquely challenging, but with detailed planning, proper advice, and solid support structures, it can be rewarding.

# CHAPTER TWO

## Business Basics

### Conversation with Scott Stalker MGySgt retired, Founder & CEO of S2 - Stalker Solutions

---

**Thomas Kelsey:** *Tell us a little about yourself, including what you did in the military and what you're doing now.*

**Scott Stalker:** I served 31 years in the Marine Corps, retiring on August 7, 2023. My career has been diverse, starting with a deployment to Mogadishu, Somalia, in 1994 and a noncombatant evacuation operation in Tirana, Albania. I've been involved in the intelligence community from early on, with multiple deployments to Iraq and Afghanistan, typical for many of us who served during that period.

I was selected as a Master Gunnery Sergeant in the Marine Corps in 2010 and became the Intel Chief for Marine Corps Special Operations Command. Later, I served as the Joint Chief of Staff J2 Senior Enlisted, then as the Senior Leader of the

Defense Intelligence Agency, the National Security Agency, and the United States Cyber Command. I finally retired from Space Command.

Post-retirement, I've taken on several board positions, including the Wounded Warrior Project, the Marine Raider Foundation, and various space, cyber, and intelligence companies. These roles range from volunteer work to paid positions. Additionally, I provide consulting services to companies in national security and leadership development.

I've also started my own company, S2 - Stalker Solutions, focused on national security, encompassing intelligence, space, cyberspace, leadership, and professional development. Because I'm not allergic to hard work, I'm a keynote speaker, registered with several speakers' bureaus, and I've just finished some work at the National Security Agency for a client.

I am preparing to launch my YouTube page, focusing on professional development, where I plan to guide people through my experiences and insights. The best way to reach me is on LinkedIn by searching Scott Stalker.

**Thomas Kelsey:** *I'm interviewing you because of your extensive military service and leadership experience. My book aims to guide active duty members in starting their businesses, whether for extra income, personal control, or post-military transition planning. I want your perspective as a leader on the challenges of managing a business while on active duty. When someone approached you about starting a business, what were your initial thoughts and advice?*

**Scott Stalker:** I'm all in when I hear someone wants to do this. And I'm excited. You become a better service member and person by taking on challenges and growing from them. But I would remind people of the Department of Defense's mission. Our military mission is to win our nation's battles and wars, to fight and win. What that means is sometimes, as leaders, we have to send individuals, whether marine or airmen, into an area where they are likely to die. And so that is not "people first," that is "mission first." I emphasize open communication about balancing business interests with military duties. Service members must understand that their business cannot interfere with their military responsibilities, regardless of the situation. It's also important for commanders to assess and approve these ventures, considering the potential risks involved.

You must also consider ethics, especially when a business may intersect with military operations or government contracts. Service members must navigate these ethical complexities carefully to avoid conflicts of interest. I chose to start my business on August 8th due to the sensitive nature of my prior role. I had significant access and influence as a combatant command senior enlisted leader. While I could have consulted a lawyer to find a way to start a defense consulting business ethically, I knew that my position made it a gray area. Defense contractors would have been eager to engage with me because of my role, but I recognized this as potentially unethical. Instead of pursuing this path while still on active duty, I decided to wait for a clean start after my service. This decision aligns with the importance of consulting with commanders and lawyers, which I wholeheartedly support, to ensure all actions are above board.

Lastly, starting a business can be financially taxing, and service members need to be prepared for these challenges. My goal is to enable our service members to succeed in their entrepreneurial endeavors through honest, upfront conversations, balancing their growth with their commitment to our mission. If one person is successful, other military members will look up to them and say, "Wow, maybe I can do this too."

**Thomas Kelsey:** *You touched on the benefits of owning a business, like personal growth through overcoming challenges. Can you elaborate on specific benefits or skills gained? Do you have examples of individuals who've successfully managed businesses and seen an increase in productivity or work quality?*

**Scott Stalker:** I've observed how running a business can enhance one's abilities numerous times. The concept of multitasking often gets a bad rap, but in business, it's about prioritizing tasks, whether they are client needs, financial obligations, or deadlines. This prioritization skill directly benefits one's performance in military roles, whether as an officer, enlisted, or civilian.

Individuals who manage businesses alongside their military duties demonstrate improved focus, time management, and effective communication in their work. Running a business requires clarity and conciseness; if you can't communicate what you're offering clearly, your clients won't understand or buy your product or service. These skills are invaluable at higher levels in the military, where senior officers expect solutions.

The professionalism, demeanor, and stress management skills gained from running a business translate to being a better service member. These qualities are essential in business and combat, whether it's presenting oneself professionally, handling stress well, or staying calm in a crisis. For instance, in the face of varying supply chain issues, remaining composed is crucial for business success, just as calmness in a crisis is an indicator of potential success in combat situations. Owning a business can significantly contribute to developing a well-rounded, efficient service member.

**Thomas Kelsey:** *Do you have any business planning tips?*

**Scott Stalker:** I firmly believe that the key to success lies in preparing the night before and laying out a strategy similar to a CONOP in the military. Dedicate 30 minutes each evening to get things in order. This can be as simple as organizing notes, reviewing engagements for the next day, and pinpointing specific talking points. It also involves practical steps like laying out your outfit, setting the alarm clock, and preparing meals. By spending these 30 minutes effectively, I've found that I'm buying myself time and ensuring I'm well-prepared.

This approach significantly reduces stress. For instance, even after a long day, when I get home at 9:15 pm, I still take the time to review everything I need for the next day. This way, I know I'll be ready for important meetings, client interactions, or to write a thank you note. Investing this half-hour each evening sets the next day up for success.

Eventually, my goal is to prepare for the next day and extend this planning to the next week and the next two weeks. I want to anticipate the needs and themes for the future.

**Thomas Kelsey:** *What resources do you recommend for an aspiring entrepreneur still on active duty?*

**Scott Stalker:** In my experience, there are three main applications I've found incredibly useful. First, the Honor Foundation was a tremendous help during my transition period. Although it's primarily a transition program, many participants, including myself, had about two years left in the military. They provided senior mentors who advised us on starting businesses, which was invaluable.

Next, I engaged with American Corporate Partners. In fact, next week, I plan to talk to my mentor from there, a CEO. He taught me everything from government contracts and invoicing to the basics like obtaining an EIN, registering an LLC, and deciding whether to go for an S corp. These were all new concepts to me, so his guidance was appreciated.

I recently spoke at an event for Boots to Business. They offer a range of support, from financial scholarships for business courses to guidance on running a successful business. They even facilitate networking with people in similar fields. For instance, if you're interested in becoming a florist, they can connect you with successful florists to discuss considerations in that field.

As a side note, whenever I find myself in a tight spot, like needing to edit a video for my YouTube channel, I often turn to YouTube itself as a resource. It's been incredibly helpful. Another great tool I use is Udemy, which offers a variety of good, free business classes. While you don't earn a certificate from these sources, in the business world, it's often more about figuring out how to do something than having a formal qualification. I rely on these platforms to learn the necessary skills and then apply them. I struggled a bit at first, but eventually, I got it done.

**Thomas Kelsey:** *I've noticed many franchise companies targeting service members, supported by data that highlights our skills in following and executing plans, managing diverse workforces, and leadership abilities – all key to making franchises successful. But I'd like to hear about your personal experience starting your own business, especially without an existing model to follow.*

**Scott Stalker:** When it comes to those franchise models, understanding the market is important. This became especially clear during COVID – how adaptable were these models to such unprecedented changes? For many businesses, not fully grasping the market dynamics and lacking financial resilience led to failure. COVID was a rare event, but similar challenges are always possible.

For me, taking the Clifton Strengths test was a game-changer. It's like Myers Briggs but zeroes in on what I'm genuinely passionate about – things that make me excited to get up in the morning, not tasks I feel obligated to do. It confirmed what I already knew about my strengths and helped me align them with my skills.

As an E9 for 13 years, I often felt restricted. There were moments when I had better ideas than my superiors, but I had to hold back. So, when I contemplated starting my own business, I knew I'd excel. On August 7th, I had everything set. By the next day, I didn't have a house or job, just my family and confidence in my network, intellect, and drive.

I hit the ground running, using LinkedIn to network and establish my brand. I never had to apply for a job – offers came to me. My advice to anyone considering a non-routine path is to follow your passions. The Clifton Strengths test showed me I needed to be in charge, capable of fast adjustments, and stay prepared. LinkedIn aided me in building a strong brand and network.

I had more opportunities than I could handle right from the start. My message is to be confident, have a plan, and know your strengths. Don't just take up something for the sake of it. After serving in the armed forces, why not do something you love? While it's hard work, it's fulfilling. I enjoy it so much that working on a Saturday doesn't bother me.

You need self-reflection, confidence, and a plan. I had been planning for two years for the day I left the Marines. I was ready for a future beyond active duty, with a strong network and understanding of the business market backed by resources and mentorship. There was no clear roadmap, so I had to learn some aspects on my own.

**Thomas Kelsey:** *In the military, we're taught to rely on our leadership for direction and purpose. But when starting a business, it's a different ballgame. One can easily spend 40 hours on platforms like LinkedIn with little results. Where did you decide to focus your efforts effectively?*

**Scott Stalker:** I've seen many approach job hunting incorrectly, appearing desperate by constantly posting job requests. I took a different route, sharing my experiences and insights in national security, space, intelligence, and mentoring. This consistent sharing built my reputation as an expert, leading to people contacting me. Just posting a resume on LinkedIn is only enough if you have a unique, in-demand skill set.

Make your name synonymous with your expertise. Your name should come to mind when someone thinks of a need in your field. I researched social media algorithms and strategies to build my brand, like learning from a LinkedIn engineer that posting twice a day isn't effective. You need to understand each platform's nuances.

I feel disheartened seeing veterans urgently post their resumes on LinkedIn. People respond to consistent, trust-building posts. Trust is what you're selling. In my case, I built trust in the national security and leadership sectors through repeated, valuable content sharing.

**Thomas Kelsey:** *Do you have any final advice for military entrepreneurs?*

**Scott Stalker:** Here's how I see it: If you never exercise or challenge your body, you're not just staying the same but actually regressing. This concept applies to your mind and personal development, too. If you stop reading, engaging with mentors, or taking on new challenges, you won't just stagnate; you'll fall behind. I always advise people to include something in their daily routine that contributes to their personal growth, whether it's a "word of the day" challenge, reading a book, or engaging in thoughtful discussions.

Neglecting self-investment means getting surpassed by others who are continuously improving. Doubling your value can double your profits, and to 10x your profits, you must 10x your value. My life's focus has been enhancing my value by spending time with successful individuals, reducing excess baggage, and minimizing distractions like excessive social media use, streaming, or biased news consumption.

The key is taking control of your time and prioritizing self-improvement, whether it's your health, finances, or education. It's not about acquiring certificates constantly but about continuous growth. Each year, I curate a reading list. People pay attention to it, often asking me about my current reading. Regardless of the genre, reading is a way to stimulate and engage your brain. If I were to emphasize one thing, it would be the importance of continually investing in yourself.

# CHAPTER THREE

## Funding Frontlines

---

No matter what kind of business you are diving into, money will be the medium to help you achieve your goals. More than considering how much money you have (or don't have), it's understanding your funding options and the reality of how investors view them. Financing is the cornerstone of starting and growing a business. But as a military entrepreneur, your path will differ from your civilian counterparts. At this crucial point, your service can have a significant benefit, and business is about focusing on strengths and mitigating weaknesses.

### *Saving Smart*

As military members, how do we build a financial nest egg to kickstart our dreams? It's all about habits and eyes on the prize. While gurus like Dave Ramsey often share a core tenet of saving, the basic principle remains the same: spend less than you earn. Consistently applying this simple rule will result in your funds growing over time, positioning you to secure targets of opportunity in the future.

### *The Tiered Approach to Borrowing Money*

If you're interested in the borrowing process, I've got you covered. The borrowing journey can be visualized as a tiered system:

➔ **Personal Savings:** It starts at home, with your savings. That's your primary seed money.

➔ **Friends & Family:** Following your personal savings, it often moves on to soft pitching your idea to friends and family. If you have a solid network and business plan, this is a viable option, but don't underestimate your obligation to repay those who believed in your journey.

➔ **Bank Loans:** After proving your business's viability to friends and family, banks are your next stop. They're looking for assurance, a solid business plan, and perhaps some collateral. Even if you don't accept investments from friends and family, the practice of talking about your business is critical before speaking with professionals.

➔ **Angel Investors & Venture Capital:** This may be a viable option if you're not keen on monthly repayments and are more open to trading equity. Remember, they're not just investing in a business but in you. Your story and journey are important.

We'll explore some of these avenues in the coming sections. But remember, every financial decision and move should align with your business vision, mission, and exit strategy.

THOMAS KELSEY

## Tapping Into Personal Savings

Especially for my fellow military members, personal savings often become the first place we think of for financing our businesses. There's a certain appeal: if you're unwilling to put your money on the line, why would someone else? It shows a level of commitment and belief in your endeavor. It is important to have communication, support, and commitment from your immediate family before you decide to spend your family's savings on a risky venture.

Using personal savings also means you retain control. You're the boss. You're not beholden to any investor or lender. No one is breathing down your neck, expecting a quick return, or pushing you in a direction you aren't ready to go in. That sense of autonomy can be empowering. Later on, when discussing business planning, you'll see how invaluable this control can be.

### Pros and Cons of Relying on Your Savings

Pros:

- **No Debt:** The best part about using your own money is you owe nothing to anyone. No interest, no monthly payments. Every dime you make goes back into growing your business or your pocket.
- **Complete Control:** Since you're not beholden to a lender or investor, every business decision is yours alone. The pace, the direction, the vision – all yours.

**Cons:**

- **Limited Leverage:** One of the challenges of solely using personal savings is its limitation on your growth potential. You might not have the vast reserves to fund aggressive expansion.
- **Personal/Family Exposure:** Since you are using personal finances to fund your venture, you are the only one focused on its success and will realize all potential losses personally.

## Loans Tailored for the Military

You might be wondering if there are loan options specifically designed for you. While military personnel often tread the same ground as civilians in the lending space, there's an ace up your sleeve: the VA. Veteran Affairs, in collaboration with the SBA, offers more favorable programs for veterans.

Once you've transitioned to the veteran side, doors begin to open. The key advantage? Education. Military veteran-centric resources often combine the funding aspect with training. It's something not commonly found in the civilian business world unless you're diving into incubators. The military-friendly private industry also supports programs like the one by USAA in partnership with the Bunker Lab that focus on equipping you with the right knowledge, ensuring that you start and flourish in your entrepreneurial journey.

However, there's a caveat for active military members. Securing loans can be a challenge. Your dual commitment to business and service can be concerning from a bank's perspective. How can they be sure you won't need to prioritize service over your business venture? This balance can create a dilemma when it comes to underwriting.

### The Double-Edged Sword of Borrowing

The world of borrowing money comes with its snares. Foremost among them? Your credit score. As a principle, the military expects us to meet our debt obligations. Failing to pay your debts can lead to stern actions, including your command stepping or losing your security clearance. In such a situation, it's not just your business that's affected but your full-time service and commitment to the nation. Remember: Every loan is a responsibility.

### The Crowdfunding Advantage

Think of crowdfunding as micro-investing. Many individuals contribute small amounts toward your business idea, often in return for a tangible product or service once you meet your funding goals. It's a way of validating your concept, proving that a market is willing to pay for your idea before diving headfirst into production. Crowdfunding can be an excellent option for us because it's a vote of confidence from potential customers. It's reassurance that there's genuine interest in your product or service before you invest heavily.

Ready to give it a shot? Begin by exploring the multitude of platforms available. Each comes with its pros and cons. Research is your ally here. Especially for us, understanding our commitments and obligations is crucial. With the unpredictability of deployments and other duties, you must know what you're signing up for.

## Angel Investors and the Military: A Match Made in Business Heaven?

Angel investors are a special group that could hold the key to your business's growth. These individuals often invest in early-stage businesses that demonstrate potential for explosive growth. I've had the privilege of being a part of an angel investing community called Venture South, which focuses on the southeast region of the United States. Angel investors look for ideas with minimal viability, often in the tech realm, and believe that a business can achieve incredible success with its investment and network.

### The Military Advantage in the Angel Investing Realm

- **Networking Power:** If there's one thing we know well, it's the strength and resilience of the military community. Our network is vast and robust, with many veterans having achieved success as entrepreneurs or holding significant positions in corporate America. This network not only provides an instant bond but also opens doors. Remember, this bond doesn't negate the need for a solid business idea. It simply creates an avenue of trust.
- **Planning and Strategy:** Our military background has instilled an unparalleled ability to plan detailed operations

in complex environments. Whether mitigating risks, setting up plans, adhering to procedures, or achieving milestones, we've been there and done that. This strategic mindset is key when presenting your business plan to angel investors. They'll question and scrutinize, but with our background, we have the foundation to showcase a well-thought-out, reliable plan.

### Mistakes to Avoid in Business Financing

The most common pitfall I've seen in military businesses is believing our own hype and jumping "all in" too quickly. Due to a lack of information or overeagerness, many military entrepreneurs commit to the first financing option they come across. Whether a loan broker or an individual bank, failing to explore a broader range of choices can lead to less-than-optimal terms. So, always try to cast a wide net. Understand your options. Be selective. Your business deserves the best deal, and you deserve the best partners.

### Success Stories in Funding

For all the aspiring business moguls out there, let me share a bit of inspiration. The military community has had its fair share of success stories, many of which have effectively leveraged funding. A standout strategy I've observed is venturing into the real estate investing sphere. For instance, I've witnessed networks among West Point graduates who've pooled resources to set up individual sub-businesses focusing on the rental market. Why has this been so successful? Primarily because real estate offers predictability. It's a tangible asset, making it straightforward to secure loans since there's palpable collateral backing these loans.

Another inspiring example was an episode of Shark Tank, where a military spouse took center stage. She had fashioned a line of unique handbags. Having already established her business successfully, she took a leap of faith, stepping onto the Shark Tank platform. And it paid off!

This story is a testament to the vast range of resources and platforms available, not just to the military members but also to their spouses. So, when you think of financing or showcasing your business, remember to think outside the box. The world is filled with opportunities waiting for the right pitch!

### Words of Wisdom

Venturing into the world of business financing, particularly when you're still serving or recently transitioned from military service, can be both exciting and treacherous. Here's some guidance to keep you grounded.

### Dollars Aren't Everything

I get it; funding feels like a lifeline for your business. But remember, there's more to business than just the money. Every dollar you accept, especially from investors, can come with strings attached.

## Control and Partnerships

Funding, particularly from angel investors or venture capital, often means giving up a slice of your business. Suddenly, you've got partners. These partners can provide valuable insights, resources, and connections. But they'll also have expectations, opinions, and a stake in your decisions. If you don't set things straight from the outset – think clear partnership agreements and legal documentation – you could be in for some challenging times.

## Pressure and Obligations

Funding through loans can put you under a magnifying glass. There's a world of difference between running a business without significant external financial pressure and having to make a hefty bank payment every month. If your business has a slow month, that bank payment can feel like a mountain.

With the right strategy and the resilient spirit we've cultivated during our military careers, the path to financing success is within reach. Keep marching forward and always look for unique opportunities tailored to your experiences.

# KEY TAKEAWAYS

- Spend less than you earn and build a financial buffer, allowing you to fund and grow your business using personal savings.

- The borrowing journey can be seen as a tiered system, from personal savings to angel investors, each with its benefits and challenges.

- Military personnel have unique advantages and hurdles in the lending space. Exploring military-specific options like VA loans can offer more favorable terms.

- Crowdfunding can effectively validate a business idea and gain initial funding, but it requires thorough research and commitment.

- While financing is essential, it's important to remember that every dollar accepted, especially from investors, might come with expectations and obligations. Always prioritize the alignment of financial decisions with your business vision and mission.

# CHAPTER FOUR

## Military Precision
## in Business Planning

Before getting into the nuts and bolts of starting and managing your business, you must understand one thing: Your business plan is at the heart of everything. I genuinely believe it's one of the most crucial things in business. It's not just a set of documents you prepare for formalities. It's the roadmap that determines how you'll drive your business forward.

### *Drawing Parallels Between Business Plans and Military Planning*

To all those who have served, the military has trained us for a life of planning and execution. Most of us are intimately familiar with a Concept of Operations or, as we usually call it, "CONOP." We're trained to craft these detailed plans to mitigate risks and achieve objectives in the most challenging circumstances. A business plan serves the same purpose in the corporate world. Your business plan is, in many ways, like a CONOP. At its core, it's a strategic blueprint – laying out how you intend to mitigate business risks while achieving your objectives.

When discussing a rock-solid business plan, we're going beyond just a simple overview of what your business might be doing. Start by thinking about your company's description: Consider the "what" *and* the "why." What's driving you? What are those values you hold dear and guide your every decision?

Then there's your marketing strategy. It won't be enough to simply "advertise here or there." You need a structured game plan. Think about it in military terms: How will you strategically position your business in the face of potential competitors? And get your target audience to take notice?

Operational plans might sound a bit dry, but in reality, they are the lifeblood of your business. It's about your day-to-day, sure, but more than that, it's how you plan to keep the ship steady, the processes you'll have in the trenches, and how you'll navigate challenges.

Speaking of challenges, who's joining you in the trenches? That's where human resources come in. This involves more extensive skill than putting a job opportunity on multiple websites. You must recognize talent, understand where they fit into your vision, and help set them up for success.

And, of course, there's financing. This is where your funds are coming from and how you allocate them. Focus on strategy, planning for the long haul, and obtaining the resources to face whatever comes your way.

And there's an element many overlook, but it's pivotal: An exit strategy.

Every military operation comes to an end. Similarly, every business should have an exit strategy. When do you intend to step away? How do you want to transition out? Whether it's selling the business, passing it on, or even closing shop – knowing your exit is as vital as starting up.

## Breaking Down the Business Plan

With so many options out there, it can be overwhelming to figure out where to start when it comes to crafting your business plan. Let me simplify it for you: The SBA provides a free mentoring service via SCORE that tops my list of recommendations. They offer invaluable mentoring with extensive online resources. I've directed many toward SCORE's business plan template. The business plan template will help you gather your thoughts, identify lacking areas, and help create your business CONOP.

And while professional guidance is necessary at some junctures, beginning with these free resources can give you a solid starting point. It means that when you seek professional advice, they have a clear foundation to refine and enhance. I recommend protecting yourself by consulting a business attorney and CPA from the start. These can seem like unnecessary expenses at the beginning of a venture, but they will keep you legal and prevent significant issues from arising.

Let's look at the key elements of a business plan.

→ **Executive Summary:** Remember, while your business plan may be detailed and dense, sometimes you need a quick snapshot. An executive summary offers a concise overview, perfect for grabbing attention.

→ **Company Design:** Using SCORE's worksheets, you can quickly pull together data on your company's structure. Is it an LLC? S Corp? This section answers those questions.

→ **Product or Service:** At the heart of your venture is what you're offering. This section highlights the unique value your company brings to the table.

→ **Marketing and Market Analysis:** Consider this as recon in military terms. You're assessing the lay of the land – from competitive data and pricing strategies to distribution methods.

→ **Operations:** This might be a tad different from the "operations" you're used to in the military. It covers the administrative side: quality control, legal structures, and hiring. It's the behind-the-scenes machine keeping your business ticking.

→ **Financing:** This is where the rubber meets the road. How will you finance your business? What about projections and capitalization? And, importantly, are you considering selling off percentages at what milestones?

→ **Exit Strategy:** This is where I feel the SCORE template could improve. Every venture should have a clear endgame. You should seek professional advice 3-5 years before selling a business to help increase potential valuations. If selling is your plan, don't go at it alone; do it while things are still improving, understanding it can take a year or more. The International Business Brokers Association (IBBA) Certified Business Intermediary (CBI) certification is something you should look for in an advisor.

## Common Pitfalls in Business Planning

During my time as a business advisor, SCORE mentor, and angel investor, I've seen passionate individuals present wildly ambitious financial projections. But let's ground ourselves in reality. It's rare to see companies grow by 200-500% in EBITDA year over year consistently. As an investor, it's a red flag. It prompts questions about the entire plan's viability.

And secondly, that exit plan. Some business owners think it consists of simply knowing when and how they'll step out, neglecting contingency planning. Case in point: A business mentor of mine passed away while in growth mode, leaving multiple locations and a family scrambling without a clear plan for the business.

Your business plan is a living testament to your vision and ambition. We have all heard stories in the news or seen our community businesses soar and falter. I can't stress enough the value of planning with precision. And given our military backgrounds, precision is something we're well-acquainted with.

## Mastering Market Research

Market research isn't just a business advantage—it's imperative. Think of it as reconnaissance before a mission. It lays out the terrain, helping you identify both opportunities and challenges. This ensures the viability of your product or service by answering these questions: Who needs what you're selling? How many competitors are in the space? And what's your edge?

In today's digital age, platforms like Google act as gateways to pools of information. A simple search can uncover competitors, industry trends, and key data points that can shape your strategy. The aim? To unearth your competitive advantages. Don't get too caught up in the price war. Focus on the value you bring to the table or your unique solution.

Depending on your business focus, there are tailored resources you can lean on. If, for instance, you're aiming to provide services to the government, there's a goldmine of resources to explore. Take North Carolina, for example. Their robust military business center, often affiliated with community colleges, is a gem. These centers help identify opportunities, create business plans, and navigate military-focused services. While they're open to all, their primary focus is to help North Carolina businesses thrive by providing services or products to the federal government.

## Fine-Tuning Forecasts and Budgets

Forecasting matters whether you're seeking a loan or wooing investors; the narrative paints a picture of your business's future, laying out expectations, potential challenges, and strategies. But there's a layer of complexity for our brothers and sisters in uniform. TDY, deployments, training exercises—the military life can be unpredictable. And it's this unpredictability that we must factor into our plans. Perhaps not every year, but it's important to consider: What if I'm called away for six months? How would the business adapt? Forecasting for a military entrepreneur means preparing for these possibilities so the business survives and thrives during these times.

Start with market research—it's your compass. You can glimpse the financial performance of similar ventures, using them as benchmarks. A great place to begin? Publicly traded companies. They're goldmines of audited financial data.

But a word of caution: Remember, these are mammoths with significant resources. As you draw comparisons, you must be realistic. If you're a small business, manage your expectations. Your profit margins might not rival these giants, but that doesn't mean you can't succeed. Use their data as a starting point, then adjust to fit your scale and resources.

Forecasting and budgeting might seem overwhelming, especially with the added challenges of military life. But with solid market research and realistic benchmarks, your budget forecasts become a powerful tool, keeping your business on a steady course, come what may.

## From Planning to Achievement

One of the most rewarding parts of my career has been witnessing the tenacity and dedication of military entrepreneurs. Their unique experiences and disciplined approach often set them up for success in the business world.

Take Media Shield, for example. This thriving enterprise is helmed by co-CEOs, one of whom started multiple businesses while serving our country on active duty. Alongside his duties, he also spearheads a local chapter for SCORE. His commitment to the business world doesn't stop there; he's successfully established multiple family-owned coffee shops. It's an awe-inspiring testament to what can be achieved with thorough planning, adept forecasting, and the ability to delegate or dive in when necessary. His accomplishments showcase his business acumen and ability to make the most of every moment.

## Keeping Your Business Plan Alive

While many entrepreneurs consider a business plan a one-time effort, it's a living document. It requires regular revisiting and adjustments. As a rule of thumb, revisit your strategic plan at least annually. This helps keep your business goals aligned with your actual trajectory and to pivot when necessary.

Additionally, any significant event in your business, be it a new partnership, an acquisition, or even personal milestones like marriage, could necessitate a review. These events might shift the direction of your business or alter your end goals. Always remember that a business plan determines where you're headed *and* how you respond to the journey's unexpected turns.

To all my fellow military entrepreneurs reading this, remember that your training, discipline, and resilience have already set you on a promising path. Harness these traits, lean on available resources, and let your business plan guide you to success.

# KEY TAKEAWAYS

- Much like a military operation relies on a Concept of Operations, a business thrives on its strategic blueprint – the business plan. Both serve to mitigate risks and achieve set objectives.

- Your business plan should go beyond just the basics. It must encompass your company's core values, marketing strategies, operational procedures, financing, human resources, and an often overlooked but crucial component - an exit strategy.

- In the realm of business, market research is similar to military reconnaissance. Before you start, it's important to understand the terrain, including recognizing competition and industry trends and identifying your unique value.

- While it's tempting to have ambitious projections, you must ground your forecasts in reality – especially with the unpredictable nature of military life. Use data from similar businesses, but adjust expectations based on your scale and resources.

- Rather than a static, one-time creation, your business plan should be dynamic, revisited, and revised regularly. Respond to changes within the business world and your personal life to align the business with its goals.

# CHAPTER FIVE

## The Benefits of Military-Inspired Team Building

———

There is a fundamental principle that everyone in the military already knows: You're only as successful as the team around you. We in the military operate as cohesive units, as teams that must rely on each other. And there's a simple reason—your team sets your standard, especially at its weakest point. This basic military reality can be your secret weapon in the business world.

### Military Entrepreneurs vs. Civilian Entrepreneurs and Team Dynamics

Civilian entrepreneurs might have the luxury of diving headfirst into their venture. They can often control more of the entire process, and by grinding long hours, they can accomplish amazing things by themselves. But for us in the military, it's not always that straightforward. Sometimes, you just won't be able to do everything by yourself, especially when unplanned challenges arise. This is where your team's significance skyrockets.

In the military, we're trained to rely on our team, our compatriots. In business, these are mentors who've been down this road before, partners who can complement your skills, vendors who offer the best resources, lenders who can inject capital when needed, and professional advisors who can steer you clear of potential pitfalls. Each of these can be equated to your fellow soldiers on the battlefield—essential to your mission's success.

A robust network and team will help you execute your business plan smoothly and provide a safety net for unforeseen obstacles. These individuals and entities help you account for and navigate the uncertainties. Leverage that. Assemble your team wisely, trust their strengths, and you'll set your business up to overcome turmoil and succeed.

### Right Person, Right Job

Whether in the military or civilian sector, hiring the right person for the right job is key. The primary questions are: Does the candidate possess attributes and values that align with your business? Can they be trained for the specific role you're hiring for? There's an added layer of complexity for those actively serving or subject to frequent relocations. Sometimes, you may need an employee to step up and assume roles outside their initial job description. Therefore, flexibility is a great character trait.

When hiring, you must understand prospective employees' goals and how they correlate with their work ethic. I'm not merely referring to whether they're hardworking or lazy. Instead, the focus should be on your leadership traits, enabling their proactive nature. Can they take the initiative, especially in your absence, whether due to deployment or other commitments? They need the autonomy and trust to handle unforeseen situations, be innovative, and be problem-solvers, even if you're only available digitally.

Several platforms and networks are tailored for businesses to find new hires. Most states offer employment services and small business centers or development hubs that can help you find potential employees. For those actively serving, military communities usually have transition programs that cater to personnel transitioning to civilian life and looking for opportunities. These resources are a pathway to finding suitable candidates.

One unique aspect worth highlighting is the military's intern or externship program during their transition phase. This allows service members to work for businesses, honing skills before their service ends. From a business owner's perspective, it offers a fresh vantage point, helping mold these individuals into community pillars. Many larger corporations leverage this initiative, training these individuals while the military covers their salary to offer full-time roles upon retirement.

Above all, understanding the culture and values you wish to establish for your enterprise should be highly prioritized. The deeper your business insight, the better your hiring decisions.

## Delegation

Delegation is often vital to efficiency, especially for military entrepreneurs juggling multiple commitments. The magic of outsourcing lies in its ability to seamlessly bridge gaps in knowledge or skills. If accounting isn't your strong suit or marketing nuances elude you, outsourcing comes to the rescue. There's a book I enjoyed by Dan Sullivan, "Who Not How," where he talks about the goal of delegation, allowing you to concentrate on your strengths. Think about it: If you value your time at $100 an hour, it only makes sense to outsource tasks that cost less than that rate. This ensures your focus remains on revenues directly contributing to business profitability.

## The Great Debate: In-House vs. Outsourcing

Things get a tad complex here. Essentially, it compares hiring an independent contractor (1099) to hiring an employee (W-2). Naturally, state laws and regulations come into play, and a seasoned business attorney will be your guiding light. Here are some factors to consider:

- **Control and Flexibility:** Employees offer more control over how tasks are executed, while outsourcing often results in less oversight.
- **Financial Structure:** With employees, you are typically looking at hourly wages or base salaries, while outsourcing could be project-based or flat-rate. This includes benefits, such as health care or an employee retirement plan.

- **Growth and Training:** An in-house team is typically cultivated, trained, and grown with your business objectives in mind. On the contrary, outsourced professionals are hired guns and experts in their fields. You're not training them; you're leveraging their existing talents.

Always align your hiring strategy with your business objectives and exit plans. For those eyeing a business sale down the line, an in-house, well-trained team is often more appealing to potential buyers. This "well-oiled machine" ensures a smooth transition for new owners.

On the flip side, should you foresee rapid scalability or contraction in your business's future, the flexibility of outsourcing shines. Contracts can be short-term, allowing for adaptability. While outsourcing offers agility, remember in-house staff warrants commitment. It's important to think about human relationships and transparency and ensure that, as a leader, you uphold the dignity of those who rally behind your business vision.

### Mastering Remote Teams

If there's one element at the heart of effective remote management, it's communication. But not just any communication—it has to be clear, concise, and tailored to the individual you're addressing. Digital resources can be incredibly helpful in refining this skill, but not without the human touch. Understanding this dynamic becomes even more important with the global shift towards remote work due to unexpected circumstances like COVID. Just as businesses now manage teams spread across the globe, think of your deployment as another such scenario. Having a well-versed team in processes and digital communication guarantees smooth operation, regardless of distance. Although the military has trained you to be a leader, managing a remote team of contractors and employees is a new skill that will take time to master.

## Crafting Your Support Network

For any entrepreneur, your business network is the lifeblood of your operations. Organizations such as professional associations, BNI groups, and chambers of commerce thrive on this principle. This isn't about surrounding yourself with like-minded individuals but more about congregating with people of shared interests. A robust network mitigates the need to experience every pitfall personally. You can also use your network to grasp the community's needs. The most important advice I can give you regarding your network is to be the "genuine you." This doesn't mean staying within your comfort zone. Another word to the wise is to avoid getting involved with the community simply looking for business; it is easy to spot people who are only there for marketing and they often leave a negative impression on others.

### Building Bridges Within and Beyond the Military Community

Active involvement is the cornerstone of effective networking. This doesn't translate to attending as many events as possible with a business-first mindset. The essence of networking lies in connecting on a genuine level—understanding people and allowing them to understand you.

Building trust and relationships is the foundation, with business opportunities evolving organically over time. Consider attending a charity event: The primary focus is supporting the cause, but business possibilities will naturally arise as you engage in things that matter to you and connect with people who value those causes.

Referrals are a testament to trust. Whether it's a brick-and-mortar store or an e-commerce platform, referrals and reviews significantly shape consumer habits. People are more inclined to spend their money where they've heard positive feedback from a trusted source.

Veterans have an array of resources and organizations tailored to their needs. But the path for active-duty entrepreneurs is less charted, which inspired this book. Military entrepreneurs should unite, collaborate, and learn collectively. A single individual may stumble and falter, but our community will overcome challenges, innovate, and thrive together.

## The Snowball Effect: A Military Entrepreneur's Triumph

An active-duty service member I know had a passion for all things automotive. As he neared the end of his military tenure, he had a vision: to build an automotive business that would cater to his community and provide a seamless transition from military life.

Starting the enterprise a few years before his service ended was a strategic move. It allowed him to approach the business without the desperation of immediate profitability, often seen as "sales breath." The military took care of his primary expenses, allowing him to grow his business patiently, aiming for an income that now surpasses his military earnings.

This is where the networking magic began. While many entrepreneurs view their business from a singular lens of serving the customer directly, they occasionally overlook the broader ecosystem that can fuel their growth. That ecosystem is the network.

He began by offering automotive services, quickly establishing a reputation for quality. As word spread, referrals poured in. But that was just the beginning. Fellow car enthusiasts within his network connected him with distributors, leading to better products at reduced costs—a win-win for him and his customers.

As the business expanded, so did his team. While initially solo, he soon found contractors, thanks to his growing network, who could handle specific jobs, allowing him to scale efficiently.

After several years, this automotive venture morphed into a self-sustaining entity. While startups often struggle to achieve profitability in their initial years, this entrepreneur defied the norm. His methodical approach and an ever-evolving network set the ball rolling, leading him to a point where he not only enjoyed financial freedom but also realized his business dream fully.

## The Power of Community and Connections

When my wife and I relocated to our new area, we were fresh fish in a big pond. All we had were our military connections. Business-wise, we were practically starting from scratch in an unfamiliar territory. We initially struggled; I won't lie. Our business faced the typical challenges: figuring out marketing, understanding the local dynamics, and getting a grip on who's who in the community.

Our breakthrough moment came when we realized that business success isn't just about numbers and sales; it's about being genuinely involved in the community. For us, this involvement ranged from participating in charities, engaging with groups that shared our values, and mentoring. We had limited financial resources to assist others in need, so we gave our time.

These avenues opened doors. The more we interacted and participated, the richer our network became. And it's this very network that started guiding us. Once we established these relationships, they became our compass. For example, imagine you're searching for a lawyer to represent your business. Your town might have several of them. But who truly specializes in business law, and who's just good with their marketing gimmicks? This is where a trusted network shines. Someone with whom you've built a genuine relationship will steer you towards the right resources.

Every community is a goldmine of opportunities and connections waiting to be unearthed. Establish genuine relationships and let the power of community guide your business journey. Your success will be measured not by profits but by the impact you have along the way.

# KEY TAKEAWAYS

- Military training emphasizes the importance of teamwork, which can be leveraged in the business world by assembling a robust and complementary team.

- When hiring, it's crucial to align values and attributes with the business vision and strategy, especially in the context of military entrepreneurs who may face frequent deployments.

- Outsourcing can be a strategic move and should be based on business objectives, growth plans, and financial considerations.

- Effective communication is at the heart of managing remote teams.

- Actively participating in and connecting with your community within and beyond the military can provide invaluable opportunities for you personally and from a business perspective.

# CHAPTER SIX

## Marketing Your Business

---

Sales are the lifeblood of any business. Without revenue, you are "investing" in a hobby. Marketing helps you build your sales pipeline. Initially, your family, friends, and personal network might help kickstart your business. However, you need effective marketing strategies to achieve sustainable growth and reach beyond your initial circle.

### *Defining a Strong Brand*

The first step is creating an identity and vision tied to your business. Consider the traditional real estate agent scenario: You might work under a big brand, like Keller Williams, RE/MAX, etc., but essentially, "you" are the business to your clients. Whether you brand "Thomas Kelsey, the sales agent" or market the company name, it matters what customers see—as a personal venture or a process-driven enterprise supported by a network.

When developing your brand, think about the purpose and vision of your business. These elements dictate whether you aim to operate as a solo entrepreneur or build a team-oriented company. If you consider selling the business in the future, being the business will lower the valuations and limit your chance of a sale. Building a business with processes and teamwork can enhance your brand's strength and market appeal.

Let's explore identity through two distinctly branded military-owned businesses: Black Rifle Coffee and FedEx. Both are veteran-founded, yet they have different branding methods that cater to their unique business models and target audiences.

Black Rifle Coffee has effectively capitalized on its military roots. Its branding is heavily tied to military culture and values, such as esprit de corps. This is evident not just in its products but also in its marketing strategies and the charities it supports. Its approach goes beyond selling coffee, as it creates a community and an identity that resonates with military personnel and enthusiasts.

On the other hand, although also started by a veteran, FedEx adopts a more subdued approach to its military heritage. The company leverages the organizational and operational skills typical of military training, focusing on efficiency and customer-first policies. It's less about overt military branding and more about utilizing the skills and disciplines learned in the service to enhance business processes and reliability.

There are no "one-size-fits-all" branding methods. I advise always aligning your brand with your values and the expectations of your target market.

## Digital Marketing Strategies

Digital marketing is an integral part of your overall strategic business plan. At the very least, your digital marketing efforts should include a website. This serves as the central hub for your brand online, where customers can learn about your offerings, your story, and how to engage with your services.

It's important to understand and implement Search Engine Optimization (SEO). SEO enhances your website's visibility in search engine results and drives organic traffic. Additionally, exploring digital advertising options such as geofencing or targeted paid ads can significantly increase your reach and attract specific market segments.

WordPress is a versatile platform for creating and managing your first website. Hiring a marketing consultant can be beneficial if you prefer to rely on an expert. A consultant can handle the complexities of digital marketing, allowing you to focus more on other aspects of your business.

Social media platforms are excellent for building brand awareness and engagement. Alongside a website, tools like SOCI can be handy for scheduling and managing ads across your social media accounts, ensuring a consistent presence without requiring constant manual input.

## A Note About Social Media and the Military

For military personnel, using social media to promote a business must be handled with awareness of professional obligations and the distinct separation of roles. Maintain a professional image and make it explicitly clear that your business activities are separate from your role in the military. You should not use your identity as a service member to endorse or support your business.

When engaging on social media, consider the content carefully, especially regarding sensitive or politically charged topics. Even posts on a business account can be perceived as reflections of your personal views, potentially affecting your military career and business reputation. Navigate these platforms tactfully so that your communications are seen as neither endorsements nor criticisms linked to your military status.

Here are some tips to ensure that your business endeavors do not conflict with your military duties, bring discredit to yourself or the military, or harm your military standing or business reputation:

- **Seek Approval:** Always ensure you have the necessary permissions for any business activity as a military member.
- **Maintain Professionalism:** Keep your social media and online presence clean and professional, avoiding controversial subjects.
- **Separate Roles:** Clearly define your military and business roles to your audience, ensuring no confusion or perceived conflict of interest.
- **Think Long-Term:** Consider the long-term implications of your business activities on your military career and business reputation.

## From Social Media to Business Empire

I want to share the success story of Austin Alexander, who successfully leveraged his military background to carve out a niche in the fitness industry. Initially, Austin began by creating engaging content on social media, specifically through challenges and competitions such as paying people to perform fitness tasks, pull-ups, or barbell squats in front of gyms. His platform, Battle Bunker, became a hit by integrating elements of military fitness culture, resonating with a broader audience.

Austin's journey began with simple content—memes, GIFs, and fitness challenges—that quickly captured viewers' attention. His effective use of digital marketing principles and his military-inspired fitness challenges helped him grow a large following. Austin transitioned from creating content to building a brand as his audience expanded. Today, he manages a thriving business with hundreds of thousands of followers across various social media platforms. He has diversified his revenue streams to include branded products, consulting services, and event hosting. This is a perfect example of how digital marketing can transform a personal passion into a lucrative business. His military ethos shaped the content and added authenticity to his brand, making it a compelling part of his business strategy.

### Expanding Your Reach

Although technically different and individually important, networking is marketing and vice versa. Networking expands the reach of your business and embeds your message into a community.

It involves engaging with individuals who can become customers, partners, or even advocates for your brand.

Reliability and action are key to building and maintaining professional relationships while on active duty. Be a person of your word and follow through on commitments. As a new business owner in a market, supporting and engaging with your community is beneficial. Whether through involvement in a non-profit that you're passionate about or supporting other local businesses, these activities help create strong relationships.

You have the advantage of leveraging both your military and civilian networks as a military member. While you may have products or services geared specifically towards the military, the broader network of civilian business leaders and owners can provide insights and guidance. They can help you overcome potential roadblocks and offer strategies that have worked in the civilian sector, which might be adaptable to your military-focused business.

Between deployments, I make time to implement my own methods for maintaining and expanding my professional network. Although I don't use a formal BNI (Business Network International) structure, I engage with a local version of this well-known networking strategy. To stay connected, I participate in community events such as the Chamber of Commerce and various charities.

However, it's important to mention that these engagements are not directly about making sales. My involvement with the Chamber of Commerce and charity organizations is primarily about contributing to the community and supporting the causes I believe in. This approach builds a foundation of trust and respect. When members of these networks understand what your business offers and see your genuine commitment to the community, they are more likely to consider your services or products when a need arises.

## Knowing Your Goals

Marketing, advertising, and public relations are distinct fields, each with its own set of experts. As a business owner, especially one with a military background, you'll find that marketing is the area where you'll receive the most solicitations. The market is saturated with countless firms eager to engage you with their services, often promising more than they can deliver.

I encourage you to have a very clear understanding of your business goals before you engage any marketing professionals. This clarity will allow you to direct their efforts effectively and avoid the common pitfalls of overspending or getting sidetracked by unnecessary services. Know what you want to achieve, outline your path forward, and then enlist the specialists to help steer your marketing towards those goals. Start with a solid understanding of your business identity and vision, and build your marketing strategies around these core principles to set the foundation for sustainable business growth and success.

# KEY TAKEAWAYS

- Emphasize branding that aligns with your values and target market, ensuring your business identity resonates with your audience.

- Incorporate effective digital marketing, including SEO and social media, to extend your reach and enhance brand visibility.

- Navigate using social media as a military member cautiously, maintaining professionalism and ensuring clear separation between your military and business roles.

- Leverage your military and civilian networks to expand your business reach and establish strong professional relationships.

- Define clear business goals before engaging with marketing professionals to focus efforts and avoid unnecessary expenses.

# CHAPTER SEVEN

## The Path to Growth

---

In business, we talk a lot about the desire for rapid growth, but we also want it to be secure and well-managed. It's important to scale up in a way that aligns with your goals and plan for the potential risks involved. When we think about sustainable growth, especially from the standpoint of a military entrepreneur, our goals might differ from typical growth-focused startups.

Achieving this type of growth lies in **planning and positioning**. I encourage you to develop a dynamic business plan that not only sets ambitious targets (your stretch goals) but also anticipates potential risks. This will equip you to handle unforeseen challenges, allowing for adjustments. In building a pathway to sustainable growth, remember that the foundational elements—people, processes, and technology—must be integrated into your planning and operational strategies to ensure they contribute positively to your business's upward trajectory. Executing your plan is more important than creating it.

## Are You Ready to Scale?

Determining when to scale your business can vary from one entrepreneur to another and is often influenced by different factors, including financial readiness and personal capacity. For many, the decision is predominantly about money. You must assess whether your business has the financial resources to support expansion. This might involve hiring additional staff, purchasing specialized equipment to enter new market segments, or increasing operational capacity.

A useful metric to look at is working capital and its industry-standard ratios. These ratios can guide you on how much money you need to keep in the bank to sustain operations and achieve the desired sales levels. A ratio higher than the industry standard might suggest that the business isn't utilizing its financial resources efficiently, indicating an opportunity to invest more in growth. A lower ratio signals that financial buffers are tight, and there might be a need to secure additional funds for flexibility and growth.

Apart from financial considerations, personal time commitment can also be an indicator. In my own experience, the decision to scale wasn't just about financial capability but also about maximizing our time. My wife and I decided to scale our business when we realized we couldn't dedicate any more hours to the business without compromising other aspects of our lives. We were becoming the bottleneck.

To address this bottleneck, we hired assistants to take over non-revenue-generating tasks, freeing us to focus on activities that directly contribute to the bottom line. This is where the book

I previously mentioned, "Who Not How," came into play. We decided on multiple approaches, including hiring a third-party virtual assistant firm specializing in our industry and a traditional assistant for tasks that we wanted to maintain a higher level of oversight for.

Each business's context will dictate the best approach to scaling, whether it's driven by financial readiness, personal capacity, or a combination of both. Recognizing and understanding these indicators can greatly influence the successful expansion of your business.

### Growth Pitfalls

One of the most prevalent issues I encounter in business brokerage and angel investing revolves around business owners excessively drawing from their companies. This practice, often motivated by the desire to minimize tax liabilities, can seriously undermine the business's value when it's time to sell or attract investors. It's understandable to want to leverage certain tax strategies, but it's crucial to recognize how these decisions appear from an outside perspective. For example, when profits are reduced by deductions that business owners classify as necessary or personal expenses (often termed "addbacks"), they may be normalized on paper. Yet, when banks or potential investors evaluate your business, they see these deductions as a reduction in profit.

This becomes significantly impactful due to the application of multiples in business valuation. If your business is valued at a multiple of its profits—say, two or three times—every dollar deducted for tax savings could effectively reduce your business's sale price or its attractiveness to investors by two to three times that amount. I advise working closely with your accountant to devise strategies that optimize tax benefits without jeopardizing the business's valuation. The goal should be to maintain clear, transparent financial records that reflect the business's true profitability. This way, when the time comes for a sale or investment round, the financials presented reflect the business's genuine value, maximizing the return from your business's profits.

Another challenge for business owners is balancing quick wins and long-term business health. At the core, every business must address the owner's needs and aspirations. This requires an understanding that while short-term gains are essential for immediate survival and momentum—such as generating initial revenue or reaching profitability—they should not come at the expense of long-term strategic development.

A practical way to maintain this balance is always to position your business as if you were preparing it for sale. This mindset encourages you to chase the quick wins that keep the business viable in the short term *and* diligently work on building a foundation to enhance the business's value over time.

## Funding Fundamentals

Funding provides the financial resources needed to scale your business and often brings valuable connections through the individuals or entities providing the capital. If your business growth strategy involves hiring specialized experts such as a seasoned engineer or salesperson, these professionals won't come cheaply and will likely exceed the costs of your initial hires. This is where funding becomes indispensable, enabling you to afford the expertise needed to elevate your business to the next level.

Securing funding typically starts close to home, often with personal savings, demonstrating your commitment and belief in your business. Next in line are family and friends, who might be more flexible and supportive but whose resources might be limited. External sources such as banks or angel investors become relevant for more substantial needs. Each source has advantages and disadvantages, as discussed in the "Funding Frontlines" chapter.

Remember that with funding comes additional responsibility. Most investors will want to see that you are using their money wisely. This traditionally means you will have an obligation to provide financial reports and a certain amount of relationship management. They are not investing in your business for it to remain stagnant; they expect it to grow and provide a return on their investment capital. Investors typically look for businesses that demonstrate frugality and proper capital management. They want assurance that you won't take a $100,000 check and squander it away on unnecessary expenditures. Investors value entrepreneurs who continue to be creative and innovative, finding cost-effective solutions even when a substantial amount of capital is at their disposal.

## Adapting Military Strategies for Business Growth

Excellent planning is necessary for success and growth in the military and business sectors. Military strategies, though seemingly distant from corporate environments, can offer insights and methodologies that are highly productive when adapted for business use.

Military operations are known for their strategic depth, including practices such as contact drills, operational risk management, troop-leading procedures, and structured training plans. When translated into business strategies, these elements can enhance your ability to plan thoroughly and execute efficiently.

- **Contact Drills and Operational Risk Management:** These practices encourage preparedness and the ability to respond swiftly and effectively to predictable challenges. In business, this translates to developing contingency plans and being ready to pivot swiftly as market conditions change.
- **Troop-Leading Procedures:** These are about clear communication, leadership, and development—ensuring every member of the team understands their role and the larger objectives. In business, this means creating detailed operational plans that guide your team and help them understand the reasoning behind tasks and goals.
- **Training Plans:** If you aim to scale your business, a critical step will be to assemble a capable team. The next task is training them to perform at their best, where military-inspired detailed planning comes into play to ensure everyone is well-prepared to accomplish assigned tasks to a set standard.

These military tactics encourage a prescriptive rather than a reactive approach, allowing business leaders to anticipate problems and equip their teams with the tools and knowledge to overcome them. Incorporating these strategies can significantly affect how your business prepares for and achieves growth.

## The FedEx Story

Frederick W. Smith, the founder of FedEx, who served with distinction in the military and earned numerous accolades, applied his "get it done" mentality—a hallmark of his military service—to the business world. He developed FedEx into a global leader in logistics and delivery services. His military background helped him face and overcome business challenges with strategy and discipline, leading to innovative solutions that stood out from the norms of the time.

## Broader Impacts: From Nike to GoDaddy

Individuals with military backgrounds also founded other major corporations like Nike, Walmart, ReMax, and GoDaddy. The founders of these companies brought with them skills like discipline, planning, and the ability to stay prepared and responsive. These qualities, often developed through basic military training, such as being meticulous with tasks as simple as rolling socks or making beds, have proven essential in navigating the complexities of running a successful business.

These leaders' military experiences instilled in them a readiness to seize opportunities and tackle challenges. Preparing ahead of time—a fundamental principle taught in basic training—enables leaders to capitalize on opportunities as they arise and navigate business obstacles more successfully.

As an active-duty member or veteran, you have a unique position as you venture into the business world. Creating, growing, and scaling a business are demanding tasks in themselves. Combined with the rigorous demands of military service, these tasks require exceptional dedication and strategic planning. The goal is to achieve a harmonious balance that allows you to thrive in your military and entrepreneurial roles without experiencing personal or emotional stress. The strategies discussed—leveraging discipline in planning, understanding the dynamics of funding, and carefully managing growth—are designed to help you strike this balance.

# KEY TAKEAWAYS

- Develop a dynamic business plan with ambitious goals and risk assessments to adapt and pivot as necessary.

- Evaluate your financial resources thoroughly before scaling, using metrics like working capital ratios to gauge financial health and readiness for expansion.

- Delegate lower-value tasks to focus on high-impact activities directly contributing to revenue generation, maximizing time and financial resources.

- Keep transparent and accurate financial documentation to support business valuation and attract potential investors or buyers.

- Adapt military planning and leadership techniques to enhance business operations and team efficiency, preparing your business to overcome challenges and sustain growth.

# CHAPTER EIGHT

## Preparing Your Business for Sale

———

Preparing your business for sale demands as much strategy and planning as its inception and growth phases. Drawing from my experience in coaching military entrepreneurs through SCORE and various mentorship programs, I aim to shed light on the importance of an exit strategy, which business owners often overlook. I can't emphasize enough to plan early and often. The plan can be tweaked along the way as things change, but the end goal remains the same to keep you moving in the right direction. If an unexpected event presents itself, you will have a plan that others can execute on your behalf. Some everyday events that push business owners to consider a sale include:

- **Retirement:** According to a survey from the International Business Brokers Association (IBBA), retirement is the number one reason for selling businesses $50 million in value or less. Retirement doesn't have to correlate with a specific age. Some owners are ready to move on, regardless of whether they have hit the magical "67" number.

- **Golden Opportunities:** Amongst businesses with a revenue of around $200,000 or less, a desire for a higher income often motivates exploring other opportunities. You may have been running a business while you were in the military, and as your military time ends, a high-paying offer from a civilian company could be too good to refuse.

- **Burnout:** The urge to move on can also come from burnout. Especially for military entrepreneurs, stresses can build up, and selling your business may feel like the only way to achieve your desperately desired work-life balance.

- **Unsolicited Offers:** Successful businesses often receive unsolicited offers. These can come from competitors trying to acquire, vendors who want to expand their offering to customers, or people simply trying to break into the industry. If your business is performing well, be ready to receive an offer anytime from investors.

- **Health Concerns:** Declining health or an unexpected diagnosis are unfortunate but real reasons for selling. You may be perfectly fine one day and unable to provide your business with the attention it needs the next. Without an exit plan, this stressful time becomes more challenging for you and your loved ones.

Military entrepreneurs have unique circumstances. Upcoming assignments requiring relocation or promotions can throw a wrench into plans. When those military-related life changes happen, you must reevaluate what's best for you in the totality of your life and your family and figure out what you need to do with your business ahead of time. The best way to do this is through an exit plan. When constructed thoughtfully, an exit plan can significantly increase your chance of sale and the selling price of your business.

Interestingly, many business owners don't have selling on their radar and need help understanding what it takes to sell a business or the associated timeline. My wife, Ashley Kelsey, CBI, is a business advisor focused on helping owners sell their businesses. When individuals reach out, they are often at the exploration phase and looking to understand better if their business is sellable and the most probable selling price (MPSP). Unfortunately, in her industry, it is common to come across individuals without an exit plan, ultimately finding themselves in a position that requires asset liquidation.

### Positioning Your Business for Success

First, consider your asking price to position your business for a favorable transition. If you price your business right, you will get more interest from potential buyers. That doesn't mean offering your business at an extreme discount. However, the multiple needs to make sense and be within the industry and market standards. Listing is not the same as selling your business.

You also must have clean books and records. Take a minute to think about your bookkeeping processes. Do you have everything tracked on a point-of-sale system? If you run a retail or a service business, an outsider looking at the company needs to understand where the money is coming in, where it's going out, and the margins in between. Separate your personal expenses from the business, as every dollar of profitability receives the benefit of your industry multiple. Yes, this means you might pay more in taxes in the short term.

Documented processes are equally important. Is there an easy way for someone to step in, pick up a manual, and duplicate the systems you have in place? Many large franchise companies provide excellent examples of this. If you get hired at McDonald's, the process to make a cheeseburger is explicitly stated and is the same every time.

As the owner of your business, you cannot be the single point of failure. You can still be involved and perform tasks. I'm not telling you to underutilize your skills. But you should have a manager who knows the business well and will stay on as you transition to the new owner. Titles don't matter here. A "Chief Operations Officer" is only as good as the knowledge in their head. They must have a high level of operational oversight beyond just their job. Make it a priority to understand your employees and their goals so you feel comfortable letting them take the reins and continuing to help the business thrive in your absence.

A part of this continuing success lies in having long-term contracts with clients. This increases business value by ensuring a repeatable and predictable income level. The new owner will feel confident knowing clients will stay during the transition and can count on X dollars from each contract. Be mindful that these larger contracts can be seen as a negative if they present concentration issues. A common concentration issue is when a large percentage of income is received from a single client.

Concentration concerns are not limited to customers; they can also be present on the supplier side. If you own a retail or service business, having multiple suppliers or wholesalers is imperative. When you limit yourself to one, and that relationship falls apart, you're stuck, greatly diminishing the value of your business. Potential buyers correlate multiple suppliers with less risk, which results in a more favorable dollar figure.

Most importantly, separate yourself from the business and take a different viewpoint. If you had the money you're asking for, would you willingly invest in the business you're selling and take on the responsibilities? For example, if you have a full-time military job and are working 80+ hours a week, running yourself into the ground for the business and not making a profit, you must be realistic regarding the asking price. How many people will willingly give you a large chunk of money from their savings or investments to work 80 hours a week to barely get by? Seeing your business for what it is can be difficult, but I encourage every business owner to ground their expectations in reality.

## Documentation Essentials

Clean financials are a top priority during a business sale and are supported by traditional bookkeeping documents, the first being your **tax return.** A new owner isn't going to audit you like the IRS, but they have a right to know what you are reporting. A potential buyer must understand any legitimate business deductions you and your accountant deem appropriate. This document becomes extremely important to ensure your finances are not intertwined between your personal and business life, a common mistake I see business owners making.

Consult an accountant (CPA) to verify you're doing everything legally by state or federal laws and what you can deduct or depreciate and what you cannot. Certain things get deducted from an accounting standpoint that don't look great on the business sales side. You must be strategic about your taxes when you're getting ready to sell, and that's why it's necessary to have taxes in order two to three years ahead of the sale.

—

### Conversation with
### Tim Krause, Chief Operating Officer, Cornerstone CPA
### and
### Michael D. Mitchell, CPA

**Thomas Kelsey:** *Introduce yourself and describe your experience working with startups and military personnel in the context of accounting and taxation.*

**Tim Krause:** We are a CPA firm that works primarily with small businesses, helping them understand their financials and how this information can assist them in their growth and decision-making. Many of our clients are U.S. Veterans who have transitioned to civilian life and started their own businesses. We offer a full range of managerial bookkeeping, tax preparation, and tax planning services, as these components work together and impact their personal lives.

**Michael D. Mitchell:** I am Michael D. Mitchell, CPA. I have been practicing accounting since January 1986. Over my career, I have helped thousands of startups and military personnel with income tax preparation and planning and served their various accounting needs.

**Thomas Kelsey:** *What are startups' most common business structures, and how do they differ?*

**Tim Krause:** This depends on the business owner's needs and is a more complex answer than it may seem. That said, most businesses start as a sole proprietorship (which could be through a state-level LLC). After this, the most common we see are partnerships that file their own tax return, and some choose to file an S-corporation election. We rarely see the benefit of a C-corporation for a small business, but as stated before, it all depends on the needs of the business.

**Michael D. Mitchell:** In NC, many new businesses start as limited liability companies (LLCs), and as they grow and mature their respective businesses, many elect to be taxed as an S-corporation. Unless an S-corporation election is filed, one-member LLCs are treated as sole proprietors, and LLCs with two or more members are treated as partnerships. Sole proprietors report their income on Form 1040, Schedule C. Partners, members, and S-corporation stockholders report income on Schedule E of their Form 1040 via K1 forms provided by their respective business entities. Social Security and

Medicare taxes (aka self-employment taxes) can be better controlled through S-corporation tax planning.

**Thomas Kelsey:** *How should a new entrepreneur decide on the most appropriate business structure for their startup?*

**Tim Krause:** Meet with a trusted advisor who will take the time to understand your goals, the specific risks of your industry, and the tax implications.

**Michael D. Mitchell:** Everyone's needs are different, and they should consult their CPA, tax attorney, or other business consultant to determine the best course of action.

**Thomas Kelsey:** *What are some specific considerations for military members when choosing a business structure?*

**Tim Krause:** They would want to consider specific buy-sell agreements about what would happen if a partner dies. Without a written succession plan, the partnership may dissolve at the time of death. Life insurance policies will often be set in place so the heirs can fulfill the agreement at the time of death.

Another thing to consider is being out of the country and making decisions, putting controls in place that would allow the business to continue to operate during your absence. In this situation, having an outside company do the books and send unbiased and thorough reports would be very beneficial.

**Michael D. Mitchell:** They should estimate the length of their business venture and gross revenue before incurring the administrative costs of establishing an LLC or corporation and the added costs for tax filings, etc. In many cases, a sole proprietorship will suffice.

**Thomas Kelsey:** *What are the key tax compliance issues startups must be aware of during their first year of operation?*

**Tim Krause:** There are many taxing agencies, so it is important to understand the tax landscape. Each industry, municipality, state, and federal agency may have specific requirements for your business. This is why it is important to connect with a trusted advisor to assist you with this process.

**Michael D. Mitchell:** If electing S-corporation status, all stockholders must be paid a reasonable salary before taking any profit distributions; however, stockholder loans can be repaid on the front end. Applying for all respective business licenses and ID numbers(federal and state), registering an LLC or corporation with their respective Secretary of State, and filing sales tax and payroll tax returns promptly are among many compliance issues to be aware of.

**Thomas Kelsey:** *How can failing to comply with tax regulations impact a new business?*

**Tim Krause:** The taxing agencies are usually willing to talk with businesses during the beginning process. However, not knowing what to do is not an excuse the agencies will typically consider once you are up and running. If you want to do it yourself, start by calling your local agency/government, who will usually assist you in the process. So first, register with the appropriate agencies, make sure you keep good records, make timely filings, and pay your bills.

**Michael D. Mitchell:** Failure to file and pay sales and/or payroll taxes, when applicable, can result in major penalties and, in extreme cases, can bring criminal charges. Failing to file income tax returns timely can also result in severe penalties.

**Thomas Kelsey:** *What are some common tax filing mistakes made by new business owners, and how can they avoid them?*

**Tim Krause:** We see businesses usually falling into one of two mistakes. First, they either overpay or underreport their taxes. This can be costly on both sides. A taxpayer has the right to pay no more than the correct amount. When they overpay, they incorrectly file, assume it is the bill, and do not want to ask questions. It is not uncommon that we see self-reporters paying twice the amount just because they did not know the tax classification at the local level.

The second common mistake is that business owners do not realize they need to register and pay. The agencies usually have very stiff penalties for under or not reporting, which can compound the burden on the business.

**Michael D. Mitchell:** Business owners should always seek the counsel of a CPA, tax attorney, or other reputable business consultant to plan their startup business and understand compliance issues properly. Taxpayers should mail all hard copies of any tax return via USPS Priority Mail, USPS Certified Mail, FedEx, or UPS to obtain a delivery receipt for the correspondence. E-filing returns is always the best option when available.

**Thomas Kelsey:** *Can you explain the concept of tax planning and why it's crucial for startup businesses?*

**Tim Krause:** The two main impacts on the business owner are the cash flow of the business and how the business profit impacts their personal taxes. Once the business is formed, an ongoing analysis of the business's financials and tax planning is beneficial to understand the future impact on your personal finances in the current year. Thus, understanding that current business decisions have tax implications before the end of the year when you can no longer change the outcome is paramount.

**Michael D. Mitchell:** Tax planning is unique for each client, depending on their specific circumstances. When done properly, tax planning can help save thousands of dollars, especially in the early years of a new business.

**Thomas Kelsey:** *What are some effective tax planning strategies for new businesses to minimize their tax liabilities?*

**Tim Krause:** First, set up a separate bank account for the business, keep good books, and document your business expenses. For example, keep a mileage log or mileage app if you use your vehicle for business. Things like this do add up, so even though they may seem small, they can be very beneficial. Once the basics are covered, since most small businesses are pass-through entities, it is also important to look at the impact of the flow of profits on your personal return.

**Michael D. Mitchell:** When appropriate, business owners can deduct their cell phone and internet bills, pay themselves rent for an office, hire family members, take retreats, pay for business-related meals, establish retirement account(s), pay for health insurance, buy a company vehicle (vehicles with a GVWR of 6,000 pounds are better for first-year depreciation options), etc.

**Thomas Kelsey:** *How often should a business review its tax plan, and what triggers should prompt a review?*

**Tim Krause:** Depending on your business's growth rate, which could warrant a quarterly review, a tax plan is normally beneficial once a year in the fall. This allows for time to look at capital investments, lump expenses, or divert spending to the next year. A trigger could be if you want to make any large changes or asset purchases; it is best to understand the tax implications of that decision so you can plan accordingly.

**Michael D. Mitchell:** Businesses should review their tax planning and projections at least every fall, but twice a year is recommended. If the business shows a substantial increase over its projected income, its tax planning and projections should be updated.

**Thomas Kelsey:** *Why must business owners, especially military ones, work with a CPA or tax professional?*

**Tim Krause:** The tax code has been amended or revised over 4,000 times over the past ten years. Having a trusted advisor trained and working with the business owner to provide financial oversight, tax compliance, and tax planning gives the business owner peace of mind. It allows them to focus on the business deliverables instead of the complexities of the tax code. The tax environment is constantly changing, and different opportunities and challenges arise.

**Michael D. Mitchell:** CPAs are monitored by a State Board, which always expects high ethical standards. Having practiced near Fort Liberty (formerly "Fort Bragg"), I have clients all over the United States and in many foreign countries due to the very personal relationships that we have formed.

**Thomas Kelsey:** *What should military entrepreneurs look for when selecting a CPA or tax advisor?*

**Tim Krause:** It is essential to have someone on your team who is knowledgeable in tax and business growth but also able to see the bigger picture. I would choose a trusted tax advisor who is willing to take the time to understand your needs, who is growth-minded, and who is more than just transactional.

**Michael D. Mitchell:** They should choose a CPA or other advisor with whom they enjoy speaking about their particular needs. A good advisor is always available to their clients via telephone, text, or email.

**Thomas Kelsey:** *How can a CPA help a business owner beyond just handling taxes?*

**Tim Krause:** First off, it is important to find a CPA who, through the CPA and their team, is more than transactional. A tax return is based on historical data; even though an experienced CPA can gather information and use the tax code so you pay no more taxes than you owe, this is limited as everything has already happened. You can do nothing to change it. So, finding a CPA willing to work with you throughout the year through financial advisory, bookkeeping services, or tax planning is vital to help you make the best decisions today for your future growth. This allows you to take full advantage of the scope of a CPA.

**Michael D. Mitchell:** A CPA can often refer attorneys, bankers, financial advisors, business brokers, etc. They can also advise on marketing strategies, cash flow management, bookkeeping, retirement and estate planning, and cost controls.

**Thomas Kelsey:** *What is one piece of advice you would give to military members starting their own business regarding financial management and tax obligations?*

**Tim Krause:** Have a CPA or trusted financial advisor on your team. There are so many changes to navigate that having someone who understands the terrain with specific knowledge and expertise can add a lot to a business builder's toolkit.

**Michael D. Mitchell:** Military members must always stay in compliance with their tax filings to avoid security clearance issues; a good CPA can help.

—

A **profit and loss statement** will indicate what your business has regarding money coming in and is another document buyers will want to see upfront in initial discussions. Early on, they will be interested in cash flow and profit, directly impacting the business's value. If you can't reproduce the income on a profit and loss statement and a tax return, there's an issue there, and it will cause a discrepancy. That doesn't mean you can't overcome it. But every discrepancy is a negative.

Another document that is always asked about is a **balance sheet**. It outlines your business's current assets and liabilities. The balance sheet helps you and your professional advisors understand the business's financial position.

As the transaction moves forward, a new owner will consider other documents:

→ **Equipment List:** Equipment lists are necessary if you have trucks, restaurant equipment, etc. Depending on the industry, new equipment, like trucks, can be more difficult to acquire than in the past. So, some of these items have more value in the current market. You should have these items listed on your tax return's depreciation schedule along with your significant equipment list.

→ **Lease:** A lease agreement will tell a new owner how long they have left in that location or if it is transferable. I've seen landlords demand hefty fees to transition the business, making selling unrealistic.

→ **Contracts:** Contracts with employees, vendors, and clients will build confidence in potential buyers.

→ **Work in Process Reports (WIP):** If you take payment upfront to render services later, a new owner will want to know how much of the job is completed. Then, they can appropriately allocate a certain percentage of the money you have already received toward future work.

### What Not to Do When Selling Your Business

Many business owners I work with choose to inform their key employees about the upcoming sale. This is a tough one because I don't condone being dishonest with your employees. But when you tell an employee that you are getting ready to leave, that creates anxiety and uncertainty, which often results in them finding another opportunity and destroying the possibility of selling your business or even the ability to continue running it at all.

Just as you want employees to stay on and committed, you can't take your eyes off the business as the owner when preparing for a sale. Owners often start daydreaming about "what's next" as they look forward to retiring on the beach and finally beginning to have some fun. But on average, selling a business takes eight to nine months. And if you start to let things slide, the value of your business will plummet. But here's a caution: the opposite can also be detrimental. Don't become so involved that the new owner will find it impossible to take your place. A delicate balance exists between keeping your focus and being the center of your business universe.

If you own a business that drums up sales quickly and renders services later, be careful about portraying unrealistic growth opportunities for the new owner. A gym is an excellent example of this. If you sell a bunch of one-year memberships and sell your business in two months at member capacity, no money will come in for the next 8 to 12 months. This can be overcome by creative deal structuring that a business advisor or attorney can assist you with.

### Proven Track Record

I advise having three solid years of high performance (net profit) and the previously mentioned documentation before looking to sell your business to help make it attractive to potential buyers and obtain the highest value. This all comes down to proper planning, hiring the right people, putting processes in place, and removing yourself from day-to-day operations as the owner.

In business, sales occasionally happen due to "flash in the night" circumstances like individual technologies or explosive growth. But it's not the norm. So, my goal is to help you plan for the norm, and of course, we'll all be excited if a unicorn appears outside of that.

The journey from inception to sale of a business, especially in the military entrepreneurial space, is a testament to your dedication and adaptability. By meticulously planning and executing your exit strategy, you create a legacy that impacts your industry and the lives you've touched.

# KEY TAKEAWAYS

- An exit strategy is essential for military entrepreneurs as a contingency plan for unexpected events and a guide to steer the business toward a profitable sale. Early and frequent planning allows for adjustments in response to changing circumstances while keeping the end goal in sight.

- The decision to sell a business can be driven by various factors such as retirement, opportunities for higher income, burnout, unsolicited offers, health concerns, or military-related life changes. Understanding these triggers helps in preparing for a timely and advantageous sale.

- Positioning your business desirably in the market involves setting a realistic asking price, maintaining clean and transparent financial records, and having well-documented operational processes. This preparation enhances the appeal of your business to potential buyers and can positively influence the selling price.

- Avoiding common pitfalls such as neglecting business operations during the sale process, failing to maintain employee morale, or misrepresenting future growth potentials is crucial. These mistakes can significantly diminish the value of your business and hinder a transition.

- Demonstrating a proven track record, ideally over three years, significantly increases the attractiveness of your business to potential buyers. Success here is characterized by stable and efficient operations, strong financial performance, and a business model that can thrive even in the owner's absence.

# CHAPTER NINE

## Selling Your Business

---

Selling your business comes with challenges and rewards, marking a significant moment in any military entrepreneur's career. Even if you don't have an immediate plan to sell, it's always good to have an exit strategy. Once you're ready to start the sale process, knowing what to expect will help you stay optimistic and prepared for anything as you execute that plan. A Certified Business Intermediary (CBI) can help you manage expectations and keep the sale moving forward.

### *Choosing the Right Business Broker*

Like most relationships in business, selecting a business broker, commonly referred to as a business advisor, comes down to finding someone who understands your values and goals and is willing to work with you to reach them. You will undoubtedly encounter several professionals with the same credentials, but you may just "jive" better with one person from the crowd. Ideally, find someone who understands why you're selling and is an International Business Brokers Association member. This is a leading organization where people willingly unite to better their craft, set rules and standards to protect their clients, and continue their education. Choose someone with a Certified Business Intermediary (CBI) designation through this organization because they have a high level of education, have completed a required number of transactions, and have ethical standards they are obligated to uphold.

### The Evolving Role of the Business Broker

Though each business broker will have a slight uniqueness in how they do things, there is a general process that we follow. Understanding this process will help you develop the right questions to ask along the way.

Any good business broker should first understand you, your business, and why you want to sell. Simultaneously, they will qualify you to determine if you are somebody they can genuinely help or if another resource would better serve you.

Once they determine you are a good match, the relationship becomes more of a marketing consultation. The business broker will work with you to create a listing that will confidentially make its way to business sales sites, generating multiple inquiries. At this point, their role changes to a "screener," verifying that potential buyers taking up your time are qualified to purchase the business, meeting your objectives, and signing the appropriate nondisclosure agreements to protect confidentiality.

As the sale progresses, the broker becomes the "middleman" or "middlewoman" between the buyer and the seller. Brokers representing the buyer and seller will work together to communicate between the two parties and defuse any of the emotions that often come with negotiations. As the seller, your broker's primary goal is to help the deal progress continuously in your best interest.

The broker's next role is "subject matter expert." If you've chosen a CBI, they will know how to work a deal. They will have attorneys to collaborate with and a keen creative sense when making transactions happen. Bumps are expected to arise during the sale process, and an experienced broker will be able to help you overcome challenges and complete the transaction.

One of the most powerful attributes of a business broker is they are a part of the business community focusing on small businesses, making them a valuable resource. Ashley is consistently helping former clients connect with professional advisors, networking groups, and learning opportunities to help them thrive. For many business brokers, their passion is small businesses, often leading them to the profession in the first place.

### Business Valuation Basics

There are three primary methods for determining the value of your business. In conjunction with the three primary methods, an endless supply of creative strategies can be utilized in specific situations. A broker can help you determine which blend of these strategies is typical in your market. This approach is focused on business sales data and may differ from what a CPA provides.

1. **Market Approach:** This method compares your business to similar businesses in your specific industry, region, and size.
2. **Income Approach:** This approach applies an industry multiple to your profits; Seller's Discretionary Earnings for Main Street businesses (generally around $5 million in annual revenue or less) or EBITDA for larger businesses.
3. **Asset Model:** The Asset Model considers physical assets, intellectual property, and, at times, goodwill. For example, a large used car dealership typically has millions of dollars of inventory on hand.

You might wonder if your valuation as a military business owner will differ from that of a civilian owner. The main scenario where a valuation might look different is if your business is geared toward the military. For example, suppose an active-duty client-facing owner tries to sell their military-focused social media company. In that case, it may have a lower value because it is more difficult for the new owner to have walked in their shoes and genuinely understand the client base.

In general, valuations will follow the approaches mentioned above or a combination thereof. The above valuations are simply a starting point, and the market will likely have different opinions. A less scientific but more practical approach to valuations is that your business is only worth what others are willing to pay.

As discussed in the last chapter, remember that delegating duties to managers is one of the most important aspects of positioning your business for a desirable valuation. This should come naturally to military business owners since they must prepare for deployment, TDYs, and PCSing. If the business can sustain itself while you're deployed, it should be able to sustain itself with a new owner.

Are you working *on* your business rather than in it? Are you developing a strategic vision for the future of your company? Do you have detailed processes and a trustworthy staff in place? Do you have a proven track record of recurring revenue? All of these will be integral parts of the valuation of your business.

## Stages of the Selling Process

The first step is preparation. Just like starting your business requires planning, so does exiting. It's important to meet with consultants regularly, often quarterly, for years before selling your business. Professionals in business sales, accounting, investments, and law should be in place before executing your sale. This advisory team can help you develop beneficial tax strategies, avoid legal pitfalls, and secure your future.

It is important to know that previous court rulings and laws govern business transactions, so you need experienced legal representation. As a business broker, I recommend retaining a business attorney specializing in transactions. They will inform you of your obligations, point out pitfalls, and advise you on legal risks. The attorney who helped with your divorce is not what I am recommending.

At some point, you must let people know your business is for sale. The next step is marketing unless you have received an unsolicited offer. However, even in this case, getting a second opinion is essential. Your business broker should reach out to other potential buyers because if you have one interested party, others will also consider your business.

Showing and screening are combined as your broker shows your business to potential buyers. The buyers gain access to information by signing the appropriate nondisclosure agreements, getting a glimpse into books and records, and having the chance to visit the physical location. Your broker will perform comprehensive screening throughout the entire process to keep buyers who aren't qualified or serious from wasting your time or poking around in your confidential information.

The deal enters the negotiation phase once a potential buyer submits an offer. Ideally, you will compare multiple offers to determine which one you will accept, counter, or decline. While comparing offers, it is important to evaluate the terms in each as it is not just about the final sales price. After an offer is accepted, there will be a due diligence period, typically 14 days to three months, depending on the size and complexity of the business. During due diligence, the buyer can ask questions and verify any information needed to make an informed decision. This could include reviewing bank statements and contracts or even speaking to key employees (it is essential to understand the risks of this ask). In business sales, everything is negotiable.

If the buyer feels confident, it's closing time. This is best handled by the transactional attorney you retained earlier in the process. Following the closing and champagne, the previously negotiated transitional period begins, which often includes a set time period where the previous owner will help the new owner learn the business. This is also when the buyer and seller communicate to staff, clients, and vendors about the transition so they are prepared and ready to support the new owner as they take the reins.

As a military entrepreneur, if you have a TDY coming up, the closing date might need to change, or you might need to get creative. If you don't plan for something like this appropriately, and you sign a legally binding contract that says you will provide X to the new owner, and you don't provide X, that can turn into a legal nightmare.

### The High Cost of Inadequate Due Diligence

We had a client who was an active-duty military member come to our office. He bought a dry cleaning business but didn't get professional advice on the front end. He paid well over the market value for the business, and the commingled finances of the previous owner didn't return the expected revenue. As an unrepresented first-time business buyer, the client's inexperience led to a lack of understanding of the financial situation of this business. Within months of this acquisition, the client came to our office in an attempt to transition out of the business, but it was worth significantly less than what he paid, like the old saying in real estate, "You make your money on the way in versus the way out." We consulted with him and advised him on reviewing his documentation and getting legal counsel to structure the transaction. Working with the previous seller and his attorney, we arrived at a mutual agreement to sell the business back to the prior owner. Even though his financial health suffered because he didn't have the advice initially, he could mitigate continuous losses with our guidance.

## Military Mindset

Military members can often be mission-focused and deal with setbacks or ambiguity by moving to the next ridgeline. But when selling your business, there is no more critical time to keep your eye on the current objectives. During the sales process, you cannot afford to be distracted. If you do and your business shows a downward trend, the sale's value and likelihood will decrease. It creates a lot of wasted time and negative ramifications for you personally. Strive to be emotionally invested in the selling process from start to finish. The journey of selling your business is as significant as the path you took to build it, and with the right approach, it can be an equally fulfilling experience.

# KEY TAKEAWAYS

- A well-chosen business broker, ideally a Certified Business Intermediary, effectively navigates the sale process, from understanding your business to closing the deal.

- Understanding the three primary methods of business valuation–Market Approach, Income Approach, and Asset Model–is essential for determining how different people may value your business.

- Regular consultation with an exit planner, accounting, and law professionals is vital for a strategic and legally sound business sale.

- Adequate due diligence will help buyers avoid overpaying and ensure a fair transaction.

- Maintaining focus and emotional investment throughout the selling process will preserve the value of your business.

# EPILOGUE

*"If you want something done, ask a busy person."*
- **Benjamin Franklin**

Thank you for spending your valuable time reading this book. Congratulations on reaching the end of this comprehensive business overview. Your dedication and commitment to learning about entrepreneurship are commendable, and I wish you the best fortune on your journey ahead.

This book was designed to get the ball rolling and provide a fundamental understanding of various business aspects. I hope you've found one or two key insights that can enhance your life. If you already possess the knowledge shared here, I urge you to give back to our community by sharing your success story and guiding others.

The cost of not getting started early is a significant one. You may miss out on your dreams and the opportunity to create a sustainable future. For me, the concern was that once retirement came along, our business would still be in start-up growth mode, lacking the sustainability to provide the necessary finances to reach our life goals. Don't let procrastination steal your future; take action now.

To simplify your path forward, here is a straightforward plan:

1. **Learn:** Master the basics, plan your journey, and launch your business
2. **Grow:** Expand in all areas of your life: military, business, and family
3. **Mentor**: Guide others to reach their goals

Something transformative happens when you stop working merely for a paycheck and gain the freedom to focus on what truly drives you. Our business has allowed my family to re-focus on our "why." Embrace this freedom and let it guide you toward fulfilling your passions and aspirations. Stay committed, stay focused, and most importantly, stay inspired. Thank you for allowing me to be a part of your journey. Here's to your success!

Warm regards,

*Thomas Kelsey*

As you look forward to a future of building your network, planning your journey, and launching your business, I am here to support you along the way. Contact me directly at **thomas@milbizguide.com or on LinkedIn at https:// www.linkedin.com/in/thomas-k-a1bb67163/**

# ABOUT THOMAS KELSEY

Thomas Kelsey is a dedicated service member and entrepreneur passionate about innovation and leadership. With over 16 years of active-duty service in the United States Air Force, Thomas has honed a unique skill set that blends military discipline with business acumen. Throughout his career, Thomas has successfully navigated the challenges of balancing a demanding military schedule with the rigors of starting and running businesses. Thomas's ability to adapt and thrive in high-pressure environments has earned him recognition in the military and business.

In addition to his entrepreneurial pursuits, Thomas is a mentor and advocate for fellow servicemembers aspiring to start, grow, or exit their businesses. He is a Certified Mentor with SCORE, where he has helped many clients by sharing his insights, strategies, and resources. Thomas is continuously evaluating new business investment opportunities as a member of Venture South.

Thomas resides in Florida with his family and continues actively serving while pursuing new business ventures.

www.ingramcontent.com/pod-product-compliance
Lightning Source LLC
Chambersburg PA
CBHW070939210326
41520CB00021B/6962